YOUR KNOWLEDGE HAS VALUE

- We will publish your bachelor's and
  master's thesis, essays and papers

- Your own eBook and book -
  sold worldwide in all relevant shops

- Earn money with each sale

Upload your text at www.GRIN.com
and publish for free

Suleiman Usman

The basic soil problems and possible solutions in agriculture

GRIN Publishing

Bibliographic information published by the German National Library:

The German National Library lists this publication in the National Bibliography;
detailed bibliographic data are available on the Internet at http://dnb.dnb.de .

Imprint:

Copyright © 2011 GRIN Verlag, Open Publishing GmbH
Print and binding: Books on Demand GmbH, Norderstedt Germany
ISBN: 978-3-640-92136-2

This book at GRIN:

http://www.grin.com/en/e-book/172288/the-basic-soil-problems-and-possible-solutions-in-agriculture

THE BASIC SOIL PROBLEMS AND POSIBLE SOLUTIONS IN AGRICULTURE

S. Usman[1]

[1]Natural Resource Institute, Plant Health and Environmental Group, University of Greenwich, Chatham Maritime, Central Avenue Kent UK ME4 4TB

Abstract

It is widely recognised that environmental problems such as soil degradation (erosion and desertification) affects many agricultural lands globally. These problems have caused soil quality decline, crop yield reduction, economic crisis, poverty, unemployment, and rural urban migration. Soil management practices are considered as the most vital and sustainable possible solution to control soil erosion and desertification. This management include use of organic manure, crop rotation, use of cover crop, intercropping, planting shelter belt and afforestation, provision of water ways, good surface drainage system, restoration of rangeland, regeneration and secondary forest, and political changes.

1.0 INTRODUCTION

The vast importance of soil in the development of various systems of agriculture and types of civilizations has long been recognized (Jenny, 1994). Soil is the basis of production in agriculture and forestry, an important component of the human environment (Zachar, 1982), and is a significant component of arid ecosystems (Russell and Greacen, 1977). Soil provides habitats for organisms (the soil fauna and micro-organisms) (Wild, 1993) and moisture and nutrients for the basic requirements of plant growth (Okigbo, 1991). Therefore, the science of soil has played, and continues to play, an important role in global studies of food production and Earth's natural resources (Hartemink, 2003). This work is on progress with the effort of many researchers and organizations through the provision of relevant information on global soil resources e.g. the development of the FAO-UNESCO soil map of the world. However, there is a need to improve the existing information on management and sustainability of soils in areas of poor research development such as Africa (e.g. Tor, 2001). This is to help improve the fertility and quality of soils of the region using management practices such as organic matter application, composting, proper irrigation systems, intercropping, and others. These activities may help increase food availability by improving soil quality, sustain soil fertility and maintain yield production (Blum, 1994; Zhao, 1995; Lal, 1997; World Bank, 2001; Mortimore and Adams, 2001; Osbahr and Allen, 2002; Pretty *et al.,* 2003).

1

To achieve the effective soil management activities, much more attention from farmers, and general concern from government, should be given to the sustainability of soil (using available and affordable management resources by farmers) and creation of good environmental policies (by government). This is because for many years soils in most part of the world, have been affected with problems related to soil quality depletion due to land degradation factors such as erosion (wind and water) (Lal, 1998), desertification as a result of poor vegetation cover, and human-induced activities of destroying forests (deforestation) (Gad and Abdel, 2000; ICLDD, 2001).

These types of problems might lead to serious damage to farmers' lands, which may include loss of organic matter and the deterioration of soil structural quality (Bradley and Thompson, 1998; UNEP, 2003) and subsequently lead to decreases in crop yields annually if not manage (e.g. Gachimbi *et al.,* 2002; GSST, 2006). These problems have been mentioned by researchers studying the soil and land degradations as principal causes of nutrient losses from soil (e.g. Stoorvogel and Smaling, 1990; Mango, 1996), as well as being the major constraint to sustainability in agricultural production (Okigbo, 1991; Gomes *et al.,* 2003; Su *et al.,* 2003). This decline in soil fertility, and increasing population pressure, has been met with by calls from some international scientists for a soil recapitalisation programme of sustainable and economic development (DFID, 2002). This programme could lead to: enhances and sustains food production, reversal of degradative trends and improvements in soil quality and soil resilience, manages water quality, and enhancement of environmental quality through sequestration of carbon and organic matter into the soil and biomass (Lal, 1995a; FAO, 2001; Lal, 2004; Lal *et al.,* 2004; Tieszen *et al.,* 2004; Farage *et al.,* 2007).

Soil and agriculture

Soil as one of the main resources of the biosphere (Holy, 1980) and important factor in the production of agricultural crops, forestry and horticulture (Okigbo, 1991), has been defined by many scholars with different views. According to USDA (2005) Soil is a natural substance comprised of solid minerals and organic matter, liquids and gasses that occur on the surface, occupies space, and is characterised by one or both of the following: horizons, or layers, that are distinguishable from the initial materials as a result of additions, losses, transfers and transformations of energy and matter or the ability to support plants in a natural environment. However, GSST (2006) define soil as unconsolidated mineral or organic materials on the immediate surface of the Earth that has been subjected to, and shows effects of, genetic and environmental factors of: climate (including water and temperature effects) and micro-organisms, conditioned by relief, acting on parent materials over a period of time. It can also refer to as a limited and irreplaceable resource and the growing degradation and loss of soil means that the expanding population in many part of the world is pressing this resource to its limits and its absence the biospheric environment of man will collapse with devastating results for humanity (Holy, 1980).

Fertile and productive soils in agriculture

Fertile soil may be defined as one which has a good supply of available plant nutrients to be drawn upon by plants throughout their growth (Govinda and Gopala, 1971). This definition describes the quality and ability of soil to provide the essential elements in adequate amounts and in proper balance for the growth of specified crop or plant in agriculture (DFID, 2002; GSST, 2006). However, Miller (1963) have the opinion that a fertile soil must have these nutrients not only in a reasonable amount or in suitable balance, but also in a way plants can take them from mineral and organic soil fractions and must be located in a climatic zone which provides moisture, light and heat sufficient for the need of plant under consideration.

This idea of fertile soil was opposed to the idea of sterile stone. Rocks differ in their fertility, where some of them are densely overgrown with lichens and micro-organisms, others are sparsely covered and others are more properly parts of one of same rock, which are not overgrown with lichens, but contain only certain microbial forms (bacteria, actinomycetes and fungi) (Glazovskaya, 1950; Krasil'nikov, 1961). Therefore, continuous and rapid rock and soil decompositions (by micro-organisms), provide a constant supply of minerals for plant growth (Hartemink, 2003), and, hence, the principal factor of soil fertility is determined by biological factors, mainly by micro-organisms (Krsli'nikov, 1961). Thus, a fertile soil is greatly depends on the amount of organic matter and the number of micro-organisms and their biodiversity in soil dynamic productivity.

Soil sustainability: meaning and importance in agriculture

Sustainability in agriculture has been defined by Conway (1985) as the ability of a system to maintain productivity in spite of larger disturbances such as repeated stress or a major perturbation (for example, the building of soil salinity or a sudden outbreak of new pests or diseases). This is largely dependent on the recycling efficiency of output *per* unit of resources input (Power *et al.,* 1997). On the other hand, sustainable soil management is vital for enhancing and sustaining the productivity of soil, food, livestock, water quality and other related land resources such as forestry (World Bank, 2001). It will help to minimise environmental impact such as desertification and soil erosion (Syers and Rimmer, 1994). Sustainable soil management can also maintain and increase agricultural output, reduce the risk of output falling and minimise the irreversible damage to the environment and land degradation (Wild, 2003). This management involves maintaining ground cover in the form of cover crops, mulching of crop residues, intercropping, listing (i.e. method of providing ridges on the surface of soil with firmer subsoil on top), using a minimum tillage system for as much of the annual season as possible, to achieve the goal of sustaining soil resources. Though the system of sustainability will vary from continent to continent and country to country (see Syers and Rimmer, 1994), four main components are noted (World Bank, 2001) (a) policy and sector work, (b) research and technology development, (c) knowledge sharing and extension, and (d) providing incentive, expenditure priorities and mode of financing. These components depend on the available and affordable resources to be use for people in every continent. For this reason, the aim of this paper was to discuss the basic soil problems and their possible solutions for future sustainable management under agricultural soil environments.

2.0 GENERAL DISCUSSION

2.1 Soil problems: soil degradation and consequences

Soil degradation is widely recognised as a serious problem and its environmental consequences will remain an important issue during the 21st century (Lal *et al.*, 2003) as well as the issues of global concern over the last few decades (Eswaran *et al.*, 2001). This has been given special prominence since the United Nations Conference on Environment and Development in 1993 (ICLDD, 2001). It is simply defined as the decline in soil quality caused through its misuse by humans, while natural factors may cause or even intensify a process that results explicitly from human action (Lal *et al.*, 2003).

Indeed, soil fertility decline has been a matter of concern ever since sedentary agriculture commenced some 10,000 years ago (Hartemink, 2003). This is because many agricultural lands in the world have been affected by soil degradation that leads to loss of soil, loss of micro/macro elements (e.g. N, P, K and Ca, Fe,), and decreases in crop yield (Jones, 1971; Holy, 1980; Okigbo, 1991; Lal, 1995b; Scherr, 1999; Eswaran *et al.*, 1999). Soil degradation affects many functions of soil (Blum, 1994) and has been considered as anthropogenic process that reduces the present and future capability of soil to support life on Earth (Oldeman *et al.*, 1991). It is also considered as a major threat to agricultural sustainability because it decreases actual and potential soil productivity (Lal, 1998).

However, Lal (1997) noted that, soil degradation occurs when soil cannot meet one of the following functions:

a) sustain biomass production and biodiversity including preservation and enhancement of gene pool (i.e. an artificial life simulation where populations of physical-based organisms evolve over time);

b) regulate water and air quality by filtering, buffering, detoxification, and regulating geochemical cycle (i.e. developmental path followed by individual elements or groups of elements in the crustal and sub-crustal zones of the Earth and on its surface)

c) preserve archaeological (scientific study of past culture), geological and astronomical records, and

d) support socio-economic structure, cultural and aesthetic values and provide engineering foundation.

After describing the functions of soil, Lal (1997) defined soil degradation as "the loss of actual or potential productivity and utility, and implies a decline in the soil's inherent capacity to produce economic goods and perform environmental regulatory function". This concept of soil degradation was differentiated from that of land degradation (Hartemink, 2003) which always embraces the degradation of the overall capacity of the land to produce economic goods and to perform environmental regulating functions (Hartemink, 2003). Although there are many definitions of land degradation (e.g. FAO/UNEP, 1983; Oldeman *et al.*, 1991), they all emphasized that the degradation of soil quality is the key factor of consideration when it came to sustainable management for agricultural production.

Types of soil degradation process

There are three types of soil degradative processes namely: physical, chemical, and biological, which through their interactive effects may lead to decline in soil quality over time (Lal, 1994; Lal *et al.*, 2003). However, important among physical process are a decline in soil structure leading to crusting, compaction, erosion, desertification, anaerobism, environmental pollution, and unsustainable use of natural resources (Eswarran *et al.*, 2001). Significant chemical processes include acidification, salinization, and fertility depletion, while biological processes include reduction in biomass carbon and decline in land biodiversity (Eswarran *et al.,* 2001). Classifications of these processes are based on the basic premise that a reduction in land agricultural productivity must occur (Blaikie and Brookfield, 1987). The development of rapid appraisal indication of these processes has been stressed by various organizations, including UNEP, DFID, FAO, and World Bank (see Conacher, 2001) as well as individual research (e.g. Lal, 1997; Eswarran *et al.,* 2001; Lal, *et al.,* 2003; Hartemink, 2003; Zhang *et al.,* 2006; Shi *et al.,* 2007). However, physical process is the most serious among the others because of it direct and indirect relationship with chemical and biological degradation processes. Example, soil erosion may change the entire soil structural and textural body. This effect could hinder soil microbial activities, change structural quality, and moved away with important chemical functions of soils.

Classes of land and soil degradation severity

As indicated earlier, land/soil degradation is one of the consequences of mismanagement of land and results frequently from a mismatch between land quality and land use (Beinroth *et al.,* 1994). Although research shows that soil/land degradation, can be clearly human-induced (e.g. Oldeman *et al.,* 1991; Eswaran *et al.,* 2001; Reich *et al.,* 2001; Lal *et al.,* 2003). However, it will not be concluded that there are no other factors that may cause soil/land degradation, because natural process also affect the land, for example, erosion by rain splash. However, human activities will remain the main contributing factor in increasing land degradation today. An assessment made on land resource stresses in Africa indicated that about 25 stress classes were defined according to the severity of their constraints (Table 2.1).

Table 1: Estimated vulnerability classes of land degradation in Africa (Reich *et al.,* 2001).

Vulnerability class	Area subject to desertification		Population affected	
	km^2	%	Millions	% of African pop.
Low	4,225,000	14.2	154.5	19.9
Moderate	4,741,000	15.9	196.1	25.3
High	3,213,000	10.8	134.8	17.4
Very high	1,466,000	4.9	22.4	2.9

Estimates of land area belonging to vulnerability classes and corresponding number of impacted population.

According to Lal *et al.*, (2003) this classification is normally based on the extent of severity using factors such as E (erosion), S (salinization), M (mining or related resource extraction activities) and V (decline in vegetative cover or quality i.e. ecological condition). However, Lal *et al.*, (2003) viewed these vulnerable areas affected by land degradation based on risk classes and finally defined them as follows:

a) Light – Land/terrain is suitable for agricultural use (including grazing), requiring only modification of management to maintain or restore full productivity: original biotic function still largely intact;

b) Moderate – Land/terrain is still suitable for agricultural use (including grazing) but productivity significantly impacted; requires full productivity and restore original biotic function;

c) Severity – Land/terrain in terms of respective degradation process is largely beyond restoration at a farm or ranch level through simple management or other on farm land treatment measures; requires more extensive treatment that may include major structural and engineering modification of the land or area-wide mitigation or restoration measures to restore productivity and biotic function;

d) Extreme – Land/terrain in terms of the respective degradation process is beyond restoration in the sense that drastic measures are required to re-establish landscape components (soil, aquatic system, vegetation) needed to have some level of agricultural productivity and biotic function.

There are of course considerable differences between countries with respect to impact on land degradation (Eswaran *et al.*, 1999). For example, Cleaver and Schreiber (1994) estimated that about 50% of Sub-Saharan agricultural land in Africa has lost its productivity due to degradation and about 80% of rangelands show signs of degradation. However, Villegas (2001) noted that the main cause of the degradation of natural resources in the rural areas globally is the fuel consumptions. It is widely accepted that deforestation is one of the most serious environmental hazard which might lead to instant soil quality degradation.

Two important forms of soil degradation

Generally, soil degradation can be viewed in two forms – soil degradation informs of erosion and soil degradation informs of desertification.

1. Soil erosion

Soil erosion can be defined as the detachment and deposition of soil particles by wind, water gravity and or by other forces from one place to another (Datta, 1986; Risse, 2001). The rate of soil loss by erosion is normally expressed in units of mass or volume/unit of time (Morgan, 1986). The word erosion is of Latin origin being derived from the verb *erodere* – to eat away (*rodere* – to gnaw), or to excavate (Zachar, 1982). The amount of soil erosion depends on the soil type, plant cover, moisture content and compaction (Robert, 1991).

Types of soil erosion

There are two types of erosion namely the geological and the accelerated types of erosion (Hudson, 1981). The geological type usually occurs under protective cover of natural vegetation (Datta, 1986). Erosion that exceeds the normal rate becomes unusually destructive and is referred to as accelerated erosion (Brady, 1990). It occurs when the process is influenced by man or animals (Holy, 1980; Hudson, 1981; Gad and Adbel, 2000; ICLDD, 2001; Risse, 2001) through activities such as inappropriate land management, over exploitation of forests and natural vegetation and overgrazing (Brady, 1990; Risse, 2001).

According to Whyte and Jacks (1986), when natural grassland, on even the gentlest slope, is mismanaged by injudicious cultivation, say, or by overgrazing, soil fertility is reduced, erosion commences and soon the amount of run-off water begins to increase. This idea suggested by Whyte and Jacks (1986) was also supported by Risse (2001) and Cummings (2006) whose demonstrates the processes of erosion in a theoretical aspect.

Beside detachment of soil particles, raindrops impacting on bare soil can also cause compaction) of the soil surface so that it can be difficult for water to infiltrate into the soil properly (e.g. Usman, 2003). This might creates more runoff and less water available to plants, which can decrease plant growth and ground cover leading to further erosion (Risse, 2001).

Soil erosion by water (rain)

Water erosion may be classified as sheet, rill, gully and stream erosions (Holy, 1980; Hudson, 1981; Datta, 1986; Brady, 1990; Robert, 1991; Risse, 2001) and the major contributing factors are rainfall intensity and run-off, soil erodibility, slope gradient and length and vegetation cover (Brady, 1990; Wall *et al.*, 2006).

Sheet erosion is characterized by the detachment and removal of soil more or less evenly over the whole affected area (Holy, 1980). The soil moves from raindrops resulting in the breakdown of soil surface structure and surface runoff (e.g. Brady, 1990); and occurs uniformly over the slope that may go unnoticed until most of the productive topsoil has been lost (Datta, 1986).

Rill erosion results when surface runoff concentrates forming small, yet well-defined, channels (Holy, 1980). These channels are called rills when they are small enough to not interfere with field machinery operation (Wall *et al.*, 2006) and also the same eroded channels known as gullies when they become a nuisance factor in normal tillage system (Risse, 2001; Wall *et al.*, 2006).

Soil erosion by wind

Wind erosion differs from water erosion. Winds and wind speeds affects larger surfaces (Holy, 1980) and occurs where there is loose sandy soil (Robert, 1991) as in the case of many world dryland areas. Wind erosion has been defined as the action of wind which dislodges the soil particles, transporting them from one place to another and deposits them there (Datta, 1986; Muhammed *et al.*, 2004). It is frequent where collectively precipitation is irregular or rare, vegetation is sparse, and cultivated soils are left bare for a relatively long period from harvest of crops in late September to sowing in late April the following year (Zhu and Chen, 1994; Li, 1998). Although most common in arid and semiarid regions, wind erosion might also occurs in humid climates as well (Brady, 1990). Soil wind erosion has been identified as an important process that affects both the surface features and the biological potential of soils (Marticorena *et al.*, 1997). It is essentially a dry-weather phenomenon and, hence, is stimulated by moisture deficiency and affects all kinds of soil materials (Brady, 1990). This is largely influenced by two processes, detachment and transportation (Hudson, 1981; Brady, 1990). These two processes can be seen as the removal of soil from around the bases of plants and the deposition of sand against obstructions such as rocks, bushes and banks (Roberts, 1991).

Hudson (1981) indicates that five different forms of wind erosion are being recognized worldwide. They are:

    a) detrusion (wearing away of rocks and soil projections by fine particles carried in suspension),

b) abrasion (movement of larger particles and bouncing along over the surface),

c) efflation (removal of particles, carried off in suspension),

d) extrusion (rolling away of large particles) and

e) effluxion (particles removal off downwind in bouncing action- i.e. saltation).

Implications of soil erosion by water and wind

There are two major implications caused by soil erosion (both in wind and water) namely: the on-site and off-site effects (Wall *et al.*, 2006). These are wide range of impacts considered to affect a number of branches of the national and international economy (Holy, 1980; Hudson, 1981). However, erosion has been recognized as a major constraint to sustainability in agricultural production (Okigbo, 1991; Lal, 1998; Adeleye *et al.*, 2003). It affects both agricultural and non-agricultural lands (Brady, 1990). For example, Brady (1990) showed that an average annual water and wind erosion loss from croplands is about 16.4 Mg/ha (7.3 tons/acre) and about 11 mg/ha will result in soil productivity (USDA, 1987).

The on-site impacts of soil erosion

Investigation (FAO/UNEP, 1983) shows that, the on-site impacts of soil erosion from both water and wind include the implications of soil erosion extend beyond the removal of valuable topsoil. Crop emergence, growth and yield are directly affected through the loss of natural nutrients and applied fertilizers with the soil (FAO/UNEP, 1983). Seeds and plants can be disturbed or completely removed from the eroded site (Wall *et al.*, 2006). Organic matter from the soil, residues and any applied manure is relatively light-weight and can be readily transported off the field, particularly during spring thaw conditions (Wall *et al.*, 2006). Soil quality, structure, stability and texture can be affected by the loss of soil (FAO/UNEP, 1983; FAO-UNESCO, 1988). The breakdown of aggregates and the removal of smaller particles or entire layers of soil or organic matter can weaken the structure and even change the texture (Wall *et al.*, 2006). Textural changes can, in turn affect the soil water-holding capacity, making it susceptible to extreme conditions such as drought (Brady, 1990).

The off-site impacts of soil erosion

On the other hand, the off-site impacts include eroded soil, deposited down slope that can inhibits or delays the emergence of seeds, bury small seedlings and necessitate replanting in the affected areas (Wall *et al.*, 2006). Sediment can be deposited down slope and can contribute to road damage (Graaff, 1993). According to Brady (1990), the annual costs of off-farm damages worldwide in 1980 were estimated by the Conservation Foundation to be between $3.2-13 billion, compared to about $2.2 billion for cropland damage.

Other implications of soil erosion

Other implications caused by erosion of entire natural resources includes deterioration of the cultivated fields; damage to forest; shortage of fodder and deterioration; increase in flood; reduce water supplies; damage to the resources such as irrigational channels, waterways and harbours; damage to the hydro-electric projects; damage to the communications; damage to fish ponds; adverse effects on public health, and social and economic consequences (Datta, 1986). Erosion may also leads to poverty increases, economic crisis, widespread of mosquitoes, unemployment, water contaminations, pest and disease widespread, and others.

2. Desertification

Desertification is the common phenomenon in world's dryland areas. The accepted definition on desertification of the time as stated during an international conference on land degradation and desertification (ICLDD, 2001) in Thailand was that of Dregne (1977) who defined it as "The impoverishment of terrestrial ecosystems that can be measured by reduced productivity of desirable plants, undesirable alterations in the biomass and the diversity of the micro and macro flora and fauna, accelerated soil deterioration, and increased hazards for human occupancy".

The high to very high desertification vulnerability classes occupied about 11.6% of the land surface worldwide (Eswaran *et al.,* 1999) leading to failure in rainfed and irrigation systems by reduction in land cover and biomass production in rangeland with an accompanying reduction in quality of feed for livestock; reduction of available woody plants for fuel and increased distances to harvest them; significant reduction in water quality; enhancement of sand and crop damage by sand-blasting and wind erosion; and increased gully and sheet erosion by torrential rain (Eswaran *et al.,* 1999).

Causes of desertification and implications

According to BBC science and environmental reports (BBC, 2000) desertification occurs due to loss of protective vegetation deforestation, overgrazing, ploughing and fire. These problems make soil vulnerable to being swept away by wind and water, causing soil to loss its structure and cohesion and becomes more easily eroded (Guardian, 2004). For example,

Mayrad *et al.* (2005) report that the root cause of desertification in many affected regions of the world is poverty. During the international conference on Land Degradation and Desertification (ICLDD, 2001) in Thailand it was reported that "the forest consumption world wide is about 3 million ha year^{-1}". This might be as a result of several factors but in Africa for example, the basic factors are poverty, population growth, poor agricultural performance and environmental degradation (Cleaver and Schreiber, 1994). This problem was considered to affect about 2.6 billion people or 44% of the world population (Eswaran *et al.*, 1999) and Cleaver and Schreiber (1994), estimates that about 50% of Sub-Saharan agricultural land presumably has lost its productivity due to degradation caused by desertification.

According to the United Nation Secretariat of the International Convention to Combat Desertification (UNCCD, 1997), the problem of desertification threatens about 30% of the Earth's total land area and already affects about 70% of its agricultural drylands, which are home to some 250 million people. At present, desertification directly affects about 3.6 billion hectares, 70% of the total dryland, or nearly one quarter of the total land of the world (one sixth of the world population) (UNEP, 2003). Furthermore, an increasing number of climatic models project that human-induced climate change may exacerbate desertification near the boundary of the Sahel over the next few decades (Patz1 *et al.*, 2005).

UNEP (2003) reported that desertification in the drylands areas of the world manifests itself through:

a) Over-exploitation and degradation of 3,333 million hectares or about 73% of the total area of rangelands which are of low potential for human and animal carrying capacity and a low population density but may be intrinsically resilient and might have considerable capacity to recuperate and regain their potential productivity if properly managed;

b) decline in fertility and soil structure leading gradually to soil loss in 216 million hectares of rainfed croplands or nearly 47% of their total area in the drylands, which constitute the most vulnerable and fragile marginal cultivable lands subjected to an increasing population pressure; and

c) degradation of 43 million hectares of irrigated croplands amounting to nearly 30% of their total area in the drylands, which usually have the highest agricultural potential and the greatest population densities when well managed.

2.2 Sustainable soil management practices as possible solutions

This management involves maintaining ground cover in the form of cover crops, mulching of crop residues, intercropping, listing (i.e. method of providing ridges on the surface of soil with firmer subsoil on top), and using a minimum tillage system for as much of the annual season as possible. This is with hope to achieve the goal of sustaining soil resources (Sullivan, 2004). Though the system of sustainability will vary from continent to continent and country to country (see Seyers and Rimmer, 1994), four main components are noted (World Bank, 2001):

a) policy and sector work;
b) research and technology development;
c) knowledge sharing and extension; and
d) providing incentive, expenditure priorities and mode of financing.

Use of manure and composting in agriculture

Manure is a mixture of dung, urine and crop residues (Mango, 1999) and contains lignin, nitrogen, phosphorus, potassium, calcium, magnesium and micronutrients such as iron and zinc (Nzuma and Murwira, 2000). Animal manure, often dry-composted with bedding and compound sweepings, is a vital factor in maintaining the physical properties of the soil and in providing chemical inputs, particularly phosphorus (Mortimore and Adams, 1999; Harris, 2001). Manure is the 'glue' that binds the soil particles together and plays an important role in preventing surface soil erosion, runuoff, leaching, and particle mass movement.

Composting is the controlled biological process of decomposition and recycling of organic material into a humus rich soil amendment known as compost (Risse, 2001). There are two basic methods for using compost in erosion control: compost blankets and compost filter berms. Compost filter berms are contoured runoff and erosion filtration methods usually used for steeper slopes with high erosive potential, while compost blankets or mats are surface applications of designated high quality composts on areas with erosive potential (Risse, 2001).

Mulching

Mulching is a technique of layering compostable materials directly on the soil to reduce erosion and improve soil quality (Robert, 2007). It is simply a protective layer of a material such as straw that is spread on top of the soil (Kowal and Kassam, 1978). Mulching builds soil, reduce soil erosion, reduce compaction from the impact of heavy rains, conserve soil moisture, reduces the need for frequent watering, provides shelter for soil organisms, drastically reduces weeding, and minimise soil degradation (NRC, 1993; AAFRDC, 2001). Mulches can either be organic such as grass clippings, straw, bark chips, and similar materials or inorganic such as stones, brick chips, and plastic (Ailin, 1995; FAO, 1997).

Intercropping

Intercropping is the practice of cultivating an additional crop between the main crops and is primary associated with sustainable agriculture for erosion control and improving soil fertility (Stainer 1984). There are at least four basic spatial arrangements used in intercropping namely (Sullivan, 2003):

a) *row intercropping:* growing two or more crops at the same time with at least one crop planted in rows,

b) *strip intercropping:* growing two or more crops together in strips wide enough to permit separate crop production using machines but close enough for the crops to interact,

c) *mixed intercropping:* growing two or more crops together in no distinct row arrangement, and

d) *relay intercropping:* planting a second crop into a standing crop at a time when the standing crop is at its reproductive stage but before harvesting.

Crop rotation

Crop rotation is the practice of growing two (or more) dissimilar type of crops on the same land in sequence (Bost, 1991). The typical example of this is growing millet, cowpea and sorghum on the same land in sequence. Most of the research carried out particularly in dryland farming indicated that crop rotation could play an important role for sustainable soil and environment, protect soil against runoff and erosion and minimise the direct impact of raindrop (e.g. Magbanua *et al.,* 1988; Torres *et al.,* 1989; IITA, 1997; Carsky *et al.,* 1999; Bationo *et al.* (2003).

Generally speaking, the use of crop rotation in maintaining soil fertility and soil erosion control may vary with region/site and the year/season. For example, In their studies of cereal legumes effects on cereal growth in the Sudano-sahelian zone of West Africa, Bagayoko *et al.* (2000) reported that the rotation effect although significant in most of the cases varied with sites and years, because some sites may need only two crops but others may need rotation of more than two crops, to ensure proper sustainable soil and environmental protection. Taking Sadore and Kouare villages as an example in this case, Bationo *et al.* (2003) reported that at Sadore whereas grain yield of pearl millet in 1998 was 1557 kg/ha in the continuous millet production, the millet rotated with cowpea yielded 1904 kg/ha; but in Kouare for the same year, sorghum grain yield increased by 50% due to rotation with cowpea in the Sudanian zone of west Africa.

Cover crops

Cover crops are crops grown to provide soil cover against erosion and enrich soil fertility (Sullivan, 2003). As the name implies, cover crops are used in covering and protecting the soil surface from wind and water erosion (Muhamman and Gungula, 2006). They can be annual, biennial or perennial herbaceous plants grown in pure or mixed stand during all or part of the growing season (Sullivan, 2003). They may also include Grasses, legumes, or other cover plants established for seasonal cover and conservation purposes such as plant shoots and roots (see Graaff, 1993). Example, the shoots cover the soil while the roots bind and stabilize the soil particles (Varhallen *et al.,* 2003). Cover crops such as cowpea was considered to provide protective cover on the land and prevent soil erosion as they can slow down water as it follows over the land (runoff); they can break the impact of a raindrop before it hits the soil, thus reducing its ability to erode (Guardian, 2004).

In the aspect of soil quality, soil function, and soil health; when cover crops in the form of green manure are incorporated into the soil, soil microbial activities may be promoted. Through cover crops nutrients held in the plant tissues are returned to the soil when the plants are dead, allowed to decay on the soil or incorporated into the soil and subsequent plants make use of the nutrients (Muhamman and Gungula, 2006). The dropping leaves from cover crops will also slowly enrich the soil and fosters the development of soil fauna like earth worms which play a prominent role in soil fertility (Anon, 2004). Cover crops take in nitrogen from the soil thereby reducing nitrogen loss through leaching which may contaminate the ground water table (Varhallen *et al.,* 2003).

In a cropping system where labour productivity is low and declining due to weeds inversion (Muhamman and Gungula, 2006), cover crops can provide an effective alternative to chemical weed control and a net savings in labour cost (Stockwell and Fisher, 1996). Some species of cover crops may be a non-host for a pest or may release toxic materials (allelopathic chemicals or entomopathogenic insecticides) that are harmful to the pest or other plants (Stevenson, 2007; Grzywacz, 2007). In addition, other materials such as plastic sheet covers can be use as protective cover against soil erosion (FAO, 1997). Research carried out in Pengyang county of Ningxia in China showed that corn yields increased by 70 percent with plastic sheet covers (Qiang,1992), while in Xinjiang the practice of irrigating over pored plastic film saved 30 to 50%t of the irrigation water and offered a yield increase of over 10% compared with furrow irrigation (Futang, 1993).

Afforestation and Shelter belt

Afforestation is the establishment of forest by natural succession or by the planting of trees on land where they did not grow formerly (Fenggi, 1993; Graaff, 1993). A shelterbelt consists of a combination of trees, shrubs and closely growing vegetation to check the force of wind, to protect against aerial and animal depredations and, to some extent, conserve soil moisture (Govinda and Gopala, 1971). Shelterbelts are planted in rows or groups of rows to provide shelter and act as a windbreak to protect crops or ornamental plants from effect of wind (Squires and Tow, 1991). A shelter belt needs to be dense at the bottom to produce a good effect against wind speed (see Govinda and Gopala, 1971). This can be established at right angles to the prevailing wind (Robert, 1991) and using the meteorological records of wind direction to decide the placement of shelterbelt is important (e.g. Konare, 1990). However, a good shelterbelt consists of 2 to 3 rows, the first row consisting of quick-growing shrubs, the second row consisting quick-growing trees and the third row of slow-growing but long standing trees (Govinda and Gopala, 1971).

Provision of waterways

It is important for all cropped fields, particularly those on sloping terrain, to provide waterways that drain away the surplus storm water from the land (Govinda and Gopala, 1971). Waterways should normally be grassed, (Graaff, 1993). However, the size and shape of such waterways depend upon the local conditions, but should be so regulated that they can carry away all surplus water (Govinda and Gopala, 1971). This can help to control the uniform movement of water over the entire farm by carrying away the excess run-off from sloping fields, as they serve to be physical barriers against environmental consequences (Graaff, 1993).

Good surface drainage system

This can be done in a way of combining ridge and furrows together with surface drainage. The ground is tilled into wide parallel ridges of the order of 10 m wide, with intervening furrows about half metre deep (Hudson, 1981). Also, soil walls can be used to bridge furrows across the slope (tied ridge) or pits along furrow bottoms to improve infiltration and reduce run-off (Guardian, 2004). Thus, the surface run-off can be moves across the ridge to the furrow and then down the furrow which is on a gradient of about 1 in 400 (Hudson, 1981). This method is particularly suitable for large areas of gently sloping land where there is need of some controlled surface drainage system (Phillips, 1963).

Restoration of rangeland

Restoration programmes are needed on degraded lands before farmers can be expected to apply sustainable agricultural practices (Graaff, 1993). Such restoration, depending on the circumstances of desertification, it might consist of some form of terracing (Hudson, 1981), conversion to perennial cropping (Hartemink, 2003) and supply of organic matter and fertilizer (see Millar, 1963), while in other cases, grasses could be useful (Graaff, 1993). According to FAO (1997) restoration of degraded steppe rangeland demonstrated that rhizomatic grass rehabilitated more quickly than tussock grass; the restoration process was of a mono-stable state; restoration dynamics corresponded generally to spatial changes along a grazing gradient.

Regeneration and secondary forest

Regeneration forests can be viewed as a type of land use in that they provide valuable goods and services to society, while preparing degraded lands for conversion to more intensive agricultural uses or alternative purposes (NRC, 1993). This process protects soils from erosion, restores the capacity of the land to retain rainfall, sequesters atmospheric carbon, and allows biological diversity to increase (NRC, 1993). According to Zhenda and Wang (1993) these measures mentioned to combat desertification should be effectively implemented through: high-level policy-making and the involvement of leading organizations at different levels; the provision of technical advice to land users from research institutes and science and technology departments; and cooperation with research institutes, local governments and local people.

Political changes

Good government policies may help to minimise or to some extent solve the problems of soil erosion and desertification. A notable example in Africa was SARCCUS (South African Regional Commission for Conservation and Utilization of the Soil) (Rowland, 1974). The United States of America environmental policies have also contributed greatly to environmental conservation and land resources management in USA (e.g. Hudson, 1981). More recently in the United Kingdom the soil action plan for England was created for the purpose of improving the standard of soil and its functions in England (SAPE, 2005). Such policies may provide a quality report to National/States governments on how to maintain and improve soil fertility in the entire region for proper sustainable development.

3.0 SUMMARY AND CONCLUSION

Environmental problem such as soil degradation (e.g. erosion and desertification) is widely considered as a serious problem and its environmental consequences will remain an important issue during the 21st century. Implications include loss of soil fertility, loss of crop yield and increase in poverty. It has been given special prominence since the United Nations Conference on Environment and Development in 1993, and is simply defined as the decline in soil quality caused through soil misuse by humans, while natural factors may cause or even intensify the process that result explicitly from human action. On the other hand, sustainable soil management practices are vital for enhancing and sustaining the productivity of soil, food, livestock, water quality and other related land resources such as forestry. It is more efficient in terms of reduced environmental impact, high risk of soil degradation and soil erosion. These management practices include use of manure, intercropping, crop rotation, shelterbelt, and planting cover crops etc. Thus, research should be directed to the following areas:

a) geo-physical assessment of soil physical environmental conditions of the different global soil types

b) Soil erosion and desertification need to be assess and study under practical demonstration for the benefit of rural people

c) geological and accelerated soil erosion need to be classify and understood dynamically in dryland areas

d) soil management practices such as crop rotation, use of cover crops, and use of organic manure and organic matter require more and more attention particularly in rural areas of Asia and Africa

With practical and qualitative data on the listed subject areas; more effective and functional management package is possible. Such effective management package should be aiming to improve soil quality, yield production, poverty alleviation, economic growth, and permanent soil and environmental protection against erosion and desertification.

ACKNOWLEDGEMENT

The soil problems and possible solutions discussed in this paper are distilled from the collective scientific knowledge in the field of soil science. A special thank is due to all authors of various text books, journals, scientific papers and reports cited in this paper.

This paper was prepared and organised by Suleiman Usman as part of his PhD reading experience under the supervision of Dr. Peter Burt, Natural Resource Institute the University of Greenwich, UK. Thanks and recognitions are extended to Dr. Peter Burt for his kindness.

REFERENCE

AAFRDC (2001) An introduction to wind erosion. Albeita Agric Food and Rural Development of Canada (AAFRDC). Ada Serafinchon, Canada. Pp 6-8.

Adeleye, E. O. Adeyemi, O. R. and Adedeji O. A. (2003) Soil degradation in Nigeria Causes and Remedies. In Adekunle, V., Okoko, E. and Adedutun S. (Eds) *Challenges of Environmental Sustainability in a Democratic Government*. Proceedings of the 11th Annual conference of Environment and Behaviour Association of Nigeria (EBAN) held at the Federal University of Technology, Akure, Ondo State, Nigeria. 26th – 27th November, 2003. Pp 78-87.

Ailin, X. (1995) Irrigation techniques over period plastic film for oasis agriculture in Xinjiang. *Agricultural Research in Arid Areas, Soil and Fertilize Institute, China, 13(1):48-89.*

Anon (Anonymous), (2004) Covering the soil to make it more fertile. In: *Spore* the Bimonthly flagship publication of Technical centre for Agriculture and Rural Cooperation. CTA – ACP – EU. Pp 4-5.

Bagayoko, M., A. Buerkert, G. Lung, A. Bationo and V. Romheld (2000) Cereal/legume rotation effects on cereal growth in Sudano-Sahelian West Africa: soil mineral nitrogen, *mycorrhizae* and nematodes. *Plant and soil 218: 103-116.*

Bationo, A., Traore, Z., Kimetu, J., Bagayoko, M., Kihara, J., Bado, V., Lompo, M., Tabo, R. and Koala, S. (2003) Cropping systems in the Sudano-sahelian zone: Implications on soil fertility management. The Tropical Soil Biology and Fertility Institute of CIAT, Nairobi, Kenya. Pp.14.

BBC (2000) Soil loss threatens food prospects. BBC News by environmental correspondent Alex Kirby. Monday, 22 May, 2000, 11:49 GMT, 12:49 UK. http://news.bbc.co.uk/1/sci/tech/758899.stm

Beinroth, F.H., Eswaran, H., Reich, P.F. and Van den Berg, E. (1994) Land related stresses in agroecosystems. In: *Stressed Ecosystems and Sustainable Agriculture,* eds. S.M. Virmani, J.C. Katyal, H. Eswaran and I.P. Abrol. New Delhi, India: Oxford and IBH.

Blakie, P. and Brookfield, H. (1987) Land degradation and society. Metheuen, London.

Blum, W.E.H., (1994) Soil resilience general approaches and definition. In: *Proceedings of the 15th International Symposia on Soil Science, vol.V2a, pp. 233–237.*

Bost, C. M. (1991) Factors affecting farm profitability in dryland farming: A systems approach (Eds) Squires, V. and Tow, P. (1991). Pp177-190.

Bradley, R. I. and Thompson, T. R. E. (1998) Sediment sources in the Llyn Tegid catchments phase 3. Unpublished Report to the Environment Agency.

Brady, N. C., (1990) The nature and properties of soils. – 10th ed. New York N.Y. Macmillan; London; Collier Macmillan, 1991. Pp 431-460

Carsky, R. J., Oyewole, B. and Tian, G. (1999) Integrated soil management for the savannah zone of West Africa: legume rotation and fertilizer N. *Journal of Nutrient Cycling in Agroecosytems, vol.55:2: (11).* Springer, Netherlands. Pp. 95-105.

Cleaver K. M. and Schreiber, A. G. (1994) Reversing the spiral, World Bank, Washington, D. C. (Cited in Federal Ministry of Environment Report, 2001).

Conacher, A. J. (2001) Land degradation. Paper selected from contributions to the sixth meeting of international geographical union commission on land degradation and desertification. Springer. Pp 17-18.

Conway, G. R. (1985) Agricultural ecology and farming system research. In Remenyi, J. V. (1988) (ed) Agricultural systems research for developing countries, Cambera, Australia. ACIAR Publication.

Cummings, D. (2006) Soil Erosion by Water. Department of Primary Industries, State of Victoria, Australia. Published on 27/11/2006 and available from http://www.dpi.vic.gov.au/dpi/nreninf.nsf/childdocs/

Datta, S. K. (1986) Soil conservation and Land management. International book Distribution, Society Ltd. Dehra Dun India pp5&18.

DFID, (2002) Soil fertility and Nutrient Management. Department for International Development of United Kingdom, edited by John Farington and Paul mundy. P1.

Dregne, H. (1977) Desertification of arid lands, *Econ. Goog.* 53(4): 322-331.

Eswaran, H., Lal R. and Reich, P. F. (2001) An overview: *"Land degradation"* International Conference on Land Degradation and Desertification, Khon Kaen, Thailand. Oxford Press, New Delhi, India.

Eswaran, H., Reich, P. and Beinroth, F. (1999) Global desertification tension zones. Selected paper from the 10th International Soil Conservation Organization Meeting held in May 24 – 29, 1999, at University and the USDA-ARS-National Soil Erosion Research Laboratory. P3.

FAO (1997) Dryland development and combating desertification: Bibliographic study of experiences in China. FAO Rome, Italy.

FAO (2001) Soil carbon sequestration for improved land management. World Soil Resources Reports 96. Food and Agriculture Organisation of the United Nations, Rome.

FAO/UNEP (1983) Guidelines for the control of soil degradations. FAO, Rome, Italy.

FAO-UNESCO (1988) Soil map of the world. Revised legend, ISRIC, Wagengen. In Hartemink, (2003) (Ed) Soil fertility decline in the tropics. Pp 52-54.

Farage, P. K., Ardö, J., Olsson, A., Rienzi, E. A., Ball, A. S. and Pretty, J. N. (2007) The potential for soil carbon sequestration in three tropical dryland farming systems of Africa and Latin America: A modelling approach. *Soil and Tillage Research, vol.94, Issue 2, June 2007:457-472.* Elsevier, B. V.

Futang, F. (1993) Deseart control and revegetation experiments on typical steppe in Hulun Buir grassland. *Grassland of China, (4):38-40.*

Gachimbi, L. N., de Jager, A., van Keulen, H., Thuranira, E. G. and Nandwa, S. M. (2002) Participatory diagnosis of soil nutrient depletion in semi-arid areas of Kenya. *Managing Africa's Soil No. 26: March 2002.*

Gad, A and Abdel, S. (2000) Study on desertification of irrigated arable lands in Egypt, *Egypt. J. Soil Sci.* 40 (2000) (3), pp. 373–384.

Glazovskaya, M. A. (1950) The effect of microrganisms on erosion of primary minerals. *Proceeding of the academy of science of the Kazakh, vol. 14:60.*

Gomes, L., Arrue, J. L., Lopez, M. V., Sterk, G., Richard, D., Gracia, R. Sabre, M., Gaudichet, A. and J.P. Frangi, J. P. (2003) Wind erosion in a Semiarid area of Spain: the WELSONS project, *Catena* 52 (2003), pp. 235–256.

Govinda R.S.V. and Gopala R.H.G (1971) Soil & crop productivity: Handbook for the use of workers in agricultural. - London: Asia Publishing House, 1971. Pp 7-17.

Graaff, Jan de (1993) Soil conservation and sustainable land use: an economic approach. - Amsterdam: Royal Tropical Institute. Pp 69-75.

Grzywacz, D. (2007) Entomopathogenic insecticide: Lecture note on Sustainable Pesticide Management (SPM). Natural Resource Institute (NRI), University of Greenwich, UK.

GSST (2006) Glossary of soil science term. SSSA- the Soil Science Society of America and ASF – the Agronomic Science Foundation, South segoe Madison, USA.

Guardian (2004) Soil Erosion as a Problem as Global Warming say Scientists. Available at: http://www.guardian.co.uk/internationalstory/0,3604,11480009,00.htm The Guardian International, Saturday, February 14, 2004

Harris, F. and Yusuf, M. A. (2001) Manure Management by Smallholder Farmers in the Kano close-settled Zone, Nigeria. *Journal of Experimental Agriculture, vol. 37: 319-332.* Cambridge University Press.

Hartemink, A. E. (2003) Soil fertility decline in the tropics with case studies on plantations / Alf. Wallingford: CABI, 2003. - pp 43-48.

Holy, M. (1980) Erosion and environment / translated (from the Czech MS.) by Jana Ondrácková. - Oxford (etc.): Pergamon, 1980. Pp 16-28.

Hudson, N.W. (1981) Soil Conservation. Bastford Academic and Educational Ltd. 4 Fitzhard Street, London. 27-30, 86.

ICLDD (2001) Second ICLDD: International Conference on Land Degradation and desertification. Kohon Kaen, Thailand. Oxford press, New Delhi, India.

IITA (1997) Farmers' perceptions of soil degradation. International Institute of Tropical Agriculture, IITA Annual report 1997.

Jenny, H. (1994) Factors of Soil Formation: A system of quantitative pedology. Foreword by Ronald Amundson University of California Berkeley. Dover publications Inc, New York. P1.

Jones, M. J. (1971) The maintenance of soil organic matter under continuous cultivations at samaru Zaria, Nigeria. *Journal of Agricultural science, 77:473-482.* Cambridge.

Konare, K. (1990) Meteorological assistance in the Sahelian region requirements and benefits. Economic and social benefits of meteorological and hydrological services, No. 733. World Meteorological Organization, Geneva. Pp144-155.

Kowal, J. M and Kassan, A. (1978) Agricultural ecology of savannah: a study of West Africa. Clarendon Press, Oxford, UK.

Krasil'nikov, N. A. (1961) Soil micro-organism and higher plants. Academy of science of the USSR, Moscow. National Science Foundation, Washington DC and Department of Agriculture, USA. Israel programme for scientific translation. P1.

Lal, R. (1997) Degradation and resilience of soils. *Philos. Trans. R. Soc. B*352, pp. 997–1008.

Lal, R., Griffin, M., Apt, J., Lave, L. and M. Morgan, M. (2004) Managing Soil Carbon, *Science* 304 (2004), p. 393.

Lal, R., Sobacki, T. M., Iivari, T. and Kimble, J. M. (2003) Soil degradation in the United State: Extent, Severity and Trends. CRC Press. Pp 20&45.

Lal, R. (ed.) (1994) Soil erosion research methods: Research methods. 2nd ed. Soil and Water Conservation Society (US). CRC Press. pp 18-24.

Lal, R. (1995a) Tillage System in the Tropics: Management options and Sustainability Implications. *FAO Soil Bulletin 71: Rome, Italy.*

Lal, R (1995b) Sustainable Management of Soil Resources in the Humid Tropic. United Nation University Press. Pp132-133.

Lal, R. (1997) Degradation and resilience of soils. *Philos. Trans. R. Soc. B*352, pp. 997–1008.

Lal, R. (1998) Soil erosion impact on agronomic productivity and environment quality. *Crit. Rev. Plant Sci.* 17, pp. 319–464. Cited by in Scopus.

Lal, R. (2004) Carbon sequestration in dryland ecosystems, *Environ. Manage.* 33 (2004), pp. 528–544.

Li, F.R., (1998) Studies on Arid Agro-Ecosystems. Shanxi Science and Technology Press, Xian (in Chinese).

Magbanua, R. D., Torrers, R. O. and Garrity, D. P. (1988) Crop residues management to sustain productivity. Paper presented at the 4th Annual Scientific Meeting of the Federation of crop scientist of the Philippines, Davoa city, Philippines, April 27-30, 1988.

Mango, N. A. R. (1996) Integrated Soil fertility management in Siaya District, Kenya. *Managing Africa's Soils No. 7.* IIED, Eilleen Higgins, Russell Press. Dryland Programme, IIED, 3 Endsleigh street London. p1.

Marticorena, B., Bergametti, G., Gillette, D. and Belnap, J. (1997) Factors   controlling threshold friction velocity in semiarid and arid area of the United States, *J. Geophysical Res.* 102 (1997) (19), pp. 23277–23287.

Mayrand, K., Paquin, M., Dionne, S. (2005) "Agricultural trade liberalization, poverty, and desertification in rural drylands: The role of UNCCD," Unisféra International Centre. http://www.unisfera.org/IMG/pdf/Unisfera_From_Boom_to_Dust_-Final.pdf   (Accessed 12/1/05).

Millar, C. A. (1963) Soil fertility. John Wiley & Sons Inc. United state of America.

Morgan, R. P. C. (1986) Soil Erosion and Conservation edited by Davidson. D. A. University of Strathclyde, Longman Scientific and technical Ltd. Hong Kong.

Mortimore M. and Adams, W. M. (1999) Working the Sahel: Environment and Society in northern Nigeria. *Routledge Research Global Environmental Change.* Taylor and Francis Group UK. Pp 30-38.

Mortimore, M. J. and Adams, W. M. (2001) Farmer adaptation, change and 'crises in the Sahel, *Global Environ. Change* 11 (2001), pp. 49–57.

Mohammed, A., Al-Quraishi, F., Dao Hu G., and Guo Chen J. (2004) Land   Nation plan of action to combat desertification. United National Environmental program (UNEP).

Muhamman, M. A. and Gungula, D. T. (2006) Cover crops in cereals based cropping systems of Northern Nigeria: Implication on sustainable production and wee management. *Journal of Sustainable Agriculture and Environment vol.2 (1):2006.*

NRC (1993) Sustainable agriculture and the environment in the humid tropics / Committee - Washington, D.C, National Academy Press, 1993. Pp 8-9.

Nzuma, J. K. and Murwira, H. K. (2000) Improving the management of manure in Zimbabwe. Managing Africa's soils No. 15. Dryland programme, IIED, Endsleigh street London.

Okigbo, B. N. (1991) Development of sustainable agricultural production systems in Africa: roles. - Ibadan: International Institute of Tropical Agriculture, 1991. Pp 26-27.

Oldeman, L. R., Hakkeling, R. T. A. and Sombrock (1991) World map of the status of human-induced soil degradation. Nairobi/Wageningen, UNEP/ISRIC GLASOD Project.

Osbahr, H., Allen, C., (2002) Soil Management at Fandou Béri, SW Niger: Ethnopedological Frameworks and Soil Fertility Management. Geoderma Special Publication on Ethnopedology.

Patz1, J., Campbell-Lendrum, D., Holloway, T., and Foley, J. (2005) "Impact of Regional Climate Change on Human Health" *Nature, Vol. 438, No. 17.*

Phillips, R. L. (1963) Surface drainage systems for farm lands: Eastern United State and Canada. *Transaction of the American Society of Agricultural Engineers, 6:313-319.*

Power, J. F., Prasad, R. and Ellis, B. G. (1997) Soil fertility management for sustainable agriculture. Technology and Industrial Arts. CRC Press. p1.

Pretty, J., Morison, J. I. L. and Hine, R. E. (2003) Reducing food poverty by increasing agricultural sustainability in developing countries, *Agric. Ecosys. Environ.* 95 (2003) (1), pp. 217–234.

Qiang, X. (1992) The important roles of plastic sheet cover in food production in southern mountain areas of Ningxia. *Agricultural Research in Arid Areas, 10(3):45-50.*

Reich, P. F., Numbem, S. T., Almaraz, R. A. and Eswaran, H. (2001) Land resource stresses and desertification in Africa. In: Bridges, E.M., I.D. Hannam, L.R. Oldeman, F.W.T. Pening de Vries, S.J. Scherr, and S. Sompatpanit (eds.). Responses to Land Degradation. Proc. 2nd. International Conference on Land Degradation and Desertification, Khon Kaen, Thailand.

Risse, M. (2001) Compost Utilization for Erosion Control. Cooperative extension service, Collage of Agricultural and Environmental Science, University of Georgia. U.S Department of Agriculture and countries of the states corporations Bulletin 1200/August, 2001.

Robert, B. R. (1991) Maintaining the Resource Base. In Squires, V. and Tow, P. (Eds) Dryland farming: A system approach. Pp146-150.

Robert, P. Stone (2007) Soil Management Specialist/OMAFRA; Neil Moore - Soil and Crop. Ministry of Agriculture Food and Rural Affairs, Ontario, Canada. Queen's Printer for Ontario

Rowland, J. W. (1974) The conservation ideal. South Africa regional commission for conservation and utilization of the soil, Pretoria, South Africa.

Russell, J. S. and Greacen E. L. (1977) Soil factors in crop production in a Semi- arid Environment. University of Queensland press in association with Australian Society of soil science incorporated, University of Queensland press. p.3.

SAPE (2005) Soil Action Plan for England: 2004 – 2006 Annual Report, June, 2005.

Scherr, S. J. (1999) Soil degradation: A threat to developing country food security by 2020. International Food Policy Research Institute. P5.

Seyers J. K. and Rimmer, D.L. (1994) Soil science and sustainable land management in the tropics / edited by J.K. - Wallingford: CAB International in association with the British Society of S, 1994. Pp 46.

Shi, P., Shimizu, H., Wang, J., Liu, L., Li, Z., Fan, Y., Yu, Y., Jia, H., Zhao, Y., Wang, and Song, Y. (2007) Land degradation and blown-san Disaster in China. National Disaster Reduction Centre of China, Ministry of Civil Affairs, Bai Guang, China.

Squires, V. and Tow, P. 1991 Dryland Farming: Systems Approach: An analysis of Dryland Agriculture in Australia. Oxford University press Australia. Pp 3-5

Steiner K G. (1984) Intercropping in tropical smallholder agriculture with special reference to West Africa. Stein, West Germany, 304pp

Stevenson, P. (2007) Botanical Pesticide: Lecture note on Sustainable Pesticide Management (SPM). Natural Resource Institute (NRI), University of Greenwich, UK.

Stockwell, C., and Fisher, L. (1996) Cover crops for sustainable agriculture in West Africa: Constraints and opportunities. A workshop organized by the International Developmental Research centre (IDRC) in collaboration with Sassakawa Global 2000, the International Institute of Tropical Agriculture (IITA), the World Bank and Ministry of Rural Development (MDR) in Cotonour, Benin Republic. 1st – 3rd October, 1996.

Stoorvogel, J.J. and Smaling, E.M.A. (1990) *Assessment of soil nutrient depletion in Sub-Saharan Africa, 1983-2000.* Report 28, Winand Centre for integrated Land, Soil and Water Research (SC-DLO), Wageningen, Netherlands.

Su, Y.Z., Zhao, H. L., T.H. Zhang, T. H. and Zhao, X.Y. (2003) Soil properties following cultivation and non-grazing of a semi-arid sandy grassland in northern China, *Soil Tillage Res.* 75 (2003), pp. 27–36.

Sullivan, P. (2003) Overview of cover crops and green manures. *ATTRA National sustainable Agricultural information services.* Fayetteville, U SA.

Sullivan, P. (2004) Sustainable soil management: soil system guide. National Sustainable Agriculture Information Service. ATTRA Publication produced by the Soil Quality Institute, Natural Resource Conservation Service. Pp1-30.

Tieszen, L. L., Tappan, G. G. and Toure, A. (2004) Sequestration of carbon in soil organic matter in Senegal: an overview. *Journal of Arid Environment, vol.59, Issue 3.* Elsevier Ltd. pp409-425.

Tor, A. B. (2001) Politics, Property and Production in the West African Sahel: Understanding Natural Resources Management. *Nordic Africa Institute (NAI).* Pp240.

Torres, R. R., Garrity, R. J., Buresh, R. K., Pandey, R. T., Bantilan, F. M. Tumacas and Montecalvo, A. (1989). Production in rice based rainfed upland cropping systems. Paper presented at the 8th Annual Scientific Meeting of the Federation of crop scientist of the Philippines, Iloilo city, Philippines, April 26-28, 1989.

UNCCD (1997) United Nation conference on desertification (UNCD), round-up, plan of action and resolution. United Nation Convention to Combat Desertification. UN, New York.

UNEP, (2003) Status of desertification and implementation of the United Nation Conference, United Nation Environmental Programme UNEP 2003.

USDA (1987) Second RCA Appraisal, Analysis of conditions and Trends, Review Draft. Washington DC, US Department of Agriculture, USA.

USDA (2005) Global soil regions map. US Department of Agriculture Natural Resources Conservation, Soil Survey Division, World Soil Resource. USDA. http://soils.usda.gov/use/worldsoils Last updated Nov. 2005.

Usman, S. (2003) Determination of Infiltration rate in Sokoto Rima flood basin. The Undergraduate research project, supervise under the guide of Head of soil Science department, Faculty of Agriculture, University press, Usman Danfodiyo University Sokoto, Nigeria.

Varhallen, A., Hayes, A., and Tailor T. (2003) Cover crops: Adaptation and use of cover crops. Ministry of agriculture and Food. Ontario, Canada.

Villegas, G. J. (2001) Factors that cause deterioration of the land in proince "Loss Andes" North Bolvian Hugh plateau. Reviewed paper for scientific content selected from the 10th International Soil Conservation Organization Meeting held in May 24 – 29, 1999, at University and the USDA-ARS-National Soil Erosion Research Laboratory. P2-3.

Wall, G., Baldwin, C. S. and Shelton, I. J. (2006) Soil erosion causes and effects. Ontario Ministry of Agriculture, Food and Rural Affairs. Queen's printer for Ontario, Canada. Pp3-6.

Whyte, R. O. and Jacks, G. V. (1982) The rape of the Earth: A world Survey of soil Erosion. Faber and Faber Ltd. 24 Russell Square London, Seventh Impression march MCMLVI, Great Bretain, Maclehose and Company Ltd. The University Press Glasshow.

Wild, A. (1993) Soils and Environment: An Introduction. Cambridge University press Trumpington Street, Cambridge New York USA. p88.

World Bank, (2001) China: Air, Land & Water – Environmental Priorities for a New Millennium. Washington, DC.

Zachar, D. (1982). Soil erosion. Development in soil science 10, forest Research Institute, zoolen, Czechoslovakia. Elsevier scientific publication company. Amsterdam, Oxford, New York. p9.

Zhang, Kefeng; Li, Xianwen; Zhou, Wenhua; Zhang, Dingxiang; Yu, Zhenrong (2006) Land resource degradation in China: Analysis of status, trends and strategy. Sapiens, *The International Journal of Sustainable Development and World Ecology, Volume 13, No.5, pp. 397-408(12).*

Zhao, Q.G., (1995) Making new contribution to development of soil science the 21st century. *Acta Pedol. Sinica* **32**, pp. 1–13 (in Chinese with English abstract).

Zhu and Chen, (1994) In: Z.D. Zhu and G.T. Chen, Editors, *Sandy Desertification in China*, Science Press, Beijing (1994) pp. 7–81 (in Chinese with English abstract).